The Pregnancy of Adam Khan

J. R. Brienza

Published by J. R. Brienza, 2020.

The Pregnancy of Adam Khan

by

J.R. Brienza

In memory of my Gram. Thank you for introducing me to The world of science fiction, thriller, horror, and noir television. Trips through past episodes taught me to ponder the future. Those late nights of fluffy blankets and hot buttered popcorn have never left me.

Chapter 1 The Intruder

Twigs and branches yielded to the weight of the tires with a muffled surrender. He cut the engine just after cresting the hill. *Best to keep things quiet,* he thought as he pulled the sports car alongside the estate's imposing stone wall. On a perfect early autumn evening like this he typically put the top down, but this time he decided it was best if he wasn't seen.

The roadster's door clicked shut and without hesitation he scaled the wall with his bare hands, jamming the toes of his sneakers into the gaps between the mortar joints. In seconds, he was resting on the wall's limestone cap. From there he could see the pool, pavilion, main house, and the garages. An electronic clicking sound warned him that a surveillance camera was turning his way. There was no time to plot his entrance now.

He jumped, thick grass carpeting his fall. He was all right. At least he thought so, but he couldn't stop to check. The dogs were already barking in the distance.

He patted at his pockets. *Where is it?* He tried not to panic as the Doberman Pincers barreled toward him. *Eureka!* The zip lock baggies were still sealed, so he opened a small corner on both. Dribbles of chicken liver blood dotted the lawn as he tossed the raw treats to the canines, plastic bag and all. *That ought to keep you busy.*

Yellow blossoms of cat's claw vine brought his eye to the second-floor balcony. Everything he wanted was there, in the master bedroom, but the treacherous vines made him think twice about climbing the trellis. So he sprinted to the front door and noticed a large, lonely parcel sitting on the steps. He rang the bell with a sweaty hand.

Through a speaker a woman's voice asked, "Who is it? What do you want?"

"Zon-Zon driver. I have a delivery for this address.

"That you, Harry?" the voice asked.

"Harry's sick today. I'm covering for him."

From the kitchen, Maria pressed a button on a wall monitor to see a man with dark hair holding a big box.

She pressed another to unlock the door, and as he entered she said, "Leave it in the vestibule, and tell Harry the housekeeper was asking 'bout him."

He placed the box near a console table flanking a pair of French carved chairs. "All set!" he shouted. "I'll send your regards to Harry." He opened and closed the door, but hid behind a chiming grandfather clock as Maria rushed out to inspect the parcel.

"More junk that man don't need!" she complained.

With her curiosity satisfied, her duties called her back to the kitchen. He slipped out from behind the clock and up the winding staircase through the double doors of the master suite. Once inside, he heard the sound of the bathroom shower hissing and surveyed his surroundings to find suitcases, along with boxes, strewn about the room.

Carefully, he closed the doors behind him and slipped across to pull out the top drawer of the bedside table, only to find ordinary things: an aspirin bottle, reading glasses, and the TV remote. There was more of the same in every drawer he checked. In short, he found nothing and lots of it.

A man's singing voice emanated from the bathroom, the exact tune unrecognizable over the hissing.

The intruder entered a large walk-in closet. It was as grand as the bedroom, paneled with carved wood and brass embellishments. Suits, ties, and other clothing of the finest quality hung from the racks.

Finally, the song was made clear. "I want to hold your haa-haa-hand. I want to hold your hand." He stopped and listened to the singing as he slowly turned the doorknob to the bathroom. As luck would have it, the click of the latch was muffled by the melody.

"What the heck, Adam?"

"Sorry, Clay," said the intruder, aka Adam. "I wanted to surprise you."

"Well, you sure did that! Almost gave me a coronary to boot! You know I missed you. C'mon, give me a little sugar."

Clay offered a soft, dripping wet kiss laced with the scent of juniper berries. "Want to join me?"

Adam's eyes scanned every muscle and sighed. "I thought you'd never ask."

Chapter 2 The Infidel

Adam went about the task of arranging the multitude of suitcases, boxes, and dirty laundry Clay accumulated while on the road. *All done!* he thought, but discovered one last designer tote with brown initials and silky leather handles.

Maria tapped on the door. "Mr. Clay, did you want me to send anything to the cleaner's?"

Adam opened the door and handed her a pile of clothes.

"Mr. Adam! What a nice surprise! When did you get here?" She walked away without expecting an answer.

The tote called to him. Conscientious as he was, he would never rest knowing that there were still items to be sorted and put in their proper place. He dug through an assortment of toiletries, hair care products, sunscreen and the like, but then found a small cardboard box. His hands shook as he opened the box to see the condoms it contained. The box was empty, save for one. Adam swallowed hard and forced himself to examine the contents of a second box. He found it filled with empty condom wrappers. His chin dropped to his chest as his heart sank.

Clay emerged from the shower without noticing anything wrong. "I'm so glad you're here, Adam. I was disappointed that you weren't around when I got home."

"I don't like to be here while you're on the road. I stay at my condo."

Clay toweled himself off and began to dress. "I know, but I want you to think of this place as your home, too. What's that tiny room of yours got on Chisholm Moors?"

"When you're gone, Clay, Chisholm Moors is just a big old mansion. It makes me feel even lonelier kicking around in this place by myself. I'm more comfortable and not as sad at home, alone."

"Well, I'm here now, and after tonight I'm all yours for the next six months."

"After tonight?"

"Did you forget? Tonight is the label's big shindig. I'm the man of the hour, remember? It's American Man's quadruple platinum party!" Clay squinted at Adam to encourage his memory.

Adam threw himself flat on the bed, arms outspread on the silk coverlet. "I did forget. All I have to wear is some casual stuff in my car."

"You have last year's tux in the closet. Maria made sure it's clean and ready to go. Besides Adam, that ol' bunch of rednecks will never know it's last year's duds."

Clay's phone came to life. Adam checked the caller ID. "It's Darcy, do you want to take that?"

"No. I swear that infernal woman never gives me a minute's peace."

Adam ignored him and answered the phone anyway. "Hi, Darcy. Clay is busy. I have you on speaker."

"Adam? Why, I... I didn't expect you." Her voice quivered.

"We're getting ready for the gala, Darcy. How may I be of service?" Adam smiled into the device.

"Um...Clay talked to you?"

"Can't say I'm thrilled," replied Adam with a shrug, "but that's the country music business. I should be used to it by now."

"Great! Hey, listen hon, the reason I called is about Clay's wardrobe."

Clay stood at the edge of the room, combing through his clothes.

"What about it? Adam asked.

"There is a selection from Jose' Aquilla. It's in the black garment bag. That's what I want him to wear. No T-shirts, sneakers, or that nasty old pair of shit-kickers that he's so fond of.

Clay groaned, dropping his favorite old boots to the floor.

"You can count on me."

"Thanks. I knew I could. Oh, and Adam?"

"Yes, Darcy?"

"I just want to tell you how proud I am of you for taking it so well."

"It's no big deal, Darcy. See you tonight."

"I'll drive. You relax," said Adam.

"Why aren't we taking the car Darcy sent?"

Adam slammed the door on his roadster and tossed the used condom box onto Clay's lap.

"You really want to do this now?" asked Clay.

"Damn it, Clay! I want you to keep your dick in your pants and quit fucking around on me!"

"Language! Geeze!" Clay winced. "You kiss your mama with that mouth? Adam, you know we can't get married."

"We've been together five years now! All of this hiding and sneaking around is bullshit."

"My fans don't know, and we're going to keep it that way! The King of Pop Country, an openly gay, married man? My label would never allow it!"

"I've heard that argument before. I'm not buying it. This isn't the Dark Ages. Your true fans will love you no matter what."

"You're wrong about them, all of them, my fans included. They'll dump me faster than a jack rabbit on a date. I'm not irreplaceable, not by a long shot. Don't even get me started on those hyper-conservative religious zealots you call family."

Adam gripped the wheel. "Take it back."

"Take what back?"

"Clay, I swear I'm not moving this car until..."

Clay sighed. "The first rule of Middle Eastern vengeance is never say anything disparaging about one's kin."

"And the consequences?"

"Okay, okay. I take it back," Clay grumbled.

Chapter 3 Tiny Bubbles

By the time Adam entered the country club's ballroom, the celebration was already in full swing. Clay was surrounded by the adoring and needy. Adam found it a curious thing, to watch his interactions with them. Longtime friend or new acquaintance, it didn't matter. Clay treated them all like they were the most important person in the room. A young man Adam didn't recognize put his hand on Clay's shoulder and whispered into his ear. Clay retrieved what appeared to be an electronic key card from his breast pocket and handed it to the youth. They exchanged knowing smiles as the young man slipped away. *Why that son of a...*

"Champagne?" A waiter wearing full western livery, complete with bolo tie, offered a silver tray of glittering bubbles to Adam.

"Ladies and gentlemen!" The crowd stirred, but was still full of noisy enjoyment. "Ladies and gentlemen, and those of you who couldn't be gentlemen if you tried, may I have your attention, please?" The room fell silent as the man known as J.D. stood at the podium. "Tonight, we're all here to celebrate an enormous achievement. Our own Clay Chisholm's 'American Man' recording has gone quadruple platinum!" Cheers, applause, and whistles went up as a spotlight searched the room. "Where is that ol' rascal? Clay? Get on up here!" Clay emerged from a corridor and navigated toward the stage. "Ladies and gentlemen, I give you Clay Chisholm, the undisputed King of Pop Country!"

Adam threw his head back and bit his lower lip, seething.

"Thank you. Thank y'all very much. I don't deserve all the fuss. I am one lucky man to be surrounded with the finest talent this business has to offer. I'd like to thank the best label out there, Big Daddy Records. J.D., take a bow. And my sound men—fellas, you rock! Last but not least, I'd like to thank my manager, Darcy Banks. Darcy, where are ya, honey?"

Darcy and Adam exchanged a nod of hello as she passed him on her way to the stage.

"Not many of y'all know this," continued Clay, "but not too long ago I started listenin' to the haters, and the nay-sayers. You know the ones. They turn up their noses, and dismiss pop country as a legit genre. Well, they had me thinkin,' maybe pop country is just a fad. Can you imagine that?" Anonymous cheers and jeers erupted in the crowd. "I didn't pay no heed, and Darcy here is the reason why. She said, 'Hon, do you, and your fans will listen.' Well, she was right, and here we all are!"

The room roared with applause.

"Aren't they a fantastic team?" interrupted J.D. "Let's hear it for Clay and Darcy!" More applause and nods followed. "And it's because of that spirit of teamwork I'd like to make a very special surprise announcement."

Clay glared at J.D. *Don't.*

"Join me in lifting a glass to the brand-spanking new Mr. and Mrs. Clay..."

Adam's Champagne flute slipped.

"...Chisholm!"

And crashed to the floor.

Chapter 4 Love Hangover

Glare from early morning sunlight bounced off the water of the swimming pool, causing Adam to shield his eyes before jumping in. He hoped the icy cold would revive him, but it didn't help. So he swam, battling the nausea that gathered strength with his own wake. Waves of dry heaves forced him out of the pool to a waiting lounge chair with white cotton pillows. He gathered a towel into a lump and rested it on his face as he allowed the sun to bake the alcohol out of his system.

"Hair of the dog?"

Clay placed a tall Bloody Mary on a table next to Adam.

Adam turned up his nose. "I hate tomato juice."

"It's a Bloody Mary, not just tomato juice. Drink. It will help."

"How did this happen?"

"Well, for starters, you never drink. You don't know how to pace yourself. The key is to have some food in your stomach, and drink lots of water." Clay pushed the beverage closer.

"I'm not asking about my hangover, I'm asking about the betrayal!"

"Betrayal? C'mon, Adam, it ain't like that."

"Oh, no? The man I love, who won't entertain the thought of marrying me, goes sneaking off to wed his business partner, and a woman no less? If that isn't betrayal, I don't know what is."

Clay took a seat at the water's edge as he composed his thoughts. "I tried to tell you, but there was never a good time."

"So this was a quick, spur of the moment thing?"

"When 'American Man' soared off the charts, J.D. threatened to pull out of my contract if I didn't shed my playboy image."

"That makes no sense." Adam rolled over, turning his back to Clay.

"Actually, it does. My demographic skews to heterosexual whites 30 years and older."

"Clay, there were plenty of opportunities to tell me that you were being pressured. You didn't say anything because you are a coward."

"Now, just hold your horses. I am no coward."

"Mr. Chisholm," interrupted Maria, "a moving van has arrived with Mrs. Chisholm's things."

"Thank you, Maria."

Adam sat up in his chair. "Wait, what? She's moving in?"

"For appearances, only. Nothing changes between you and me."

Adam stood and threw the Bloody Mary at Clay. "You're wrong! This changes everything."

Chapter 5 Cry Baby

"Adam, pick up the phone. I know you're just standing there."

Adam waited on the other end as Clay groveled.

"Fine. I see how it is. If you won't come home, I'm comin' there!"

"Pfft." Adam rolled his eyes and left his condo. "Does he think I'm just going to sit around and wait for him when I have a Mid-Atlantic tour to pack for? I've got places to go and people to see."

The sports car's tires smoked on the pavement as he drove out of sight.

Later, when Adam arrived back at his condo, he found a note taped to his door. It read: *Please come home. I miss you.* Adam crumpled the note, stuffed it into his pocket, and hurried inside to pack for his tour.

Signed on as a dancer with his favorite diva's tour, Adam needed clothes for the next few weeks. He ironed, and packed his suitcase with layers of acid free tissue between his garments. His shoes were housed in individual felt bags so they wouldn't touch. Fewer things irritated Adam more than scuffed shoes. "I need that duffel. My shoes don't fit in anything else," he said to himself. Pausing to think for a minute, he remembered the note stuffed in his pocket, and exactly where he left the bag. "Damn it!"

I'm only here for my bag. Adam used his access code at the gate and drove the roadster to the crest of the governor's drive of Chisholm Moors. He let himself in through the front door of Clay's home and made his way to the master bedroom, where he retrieved his duffel bag before voices from down the hall made him fear an encounter with Clay.

"Darlin', you got skills," said Clay.

He bolted toward the stairs but came face to face with Clay, who exited a room whilst zipping up his fly.

Darcy stood in the doorway, wearing a sheer robe fastened loosely at the waist.

Clay reached for Adam's arm. "I can explain."

"I thought he knew," Darcy stated as she smoothed her mess of hair.

Adam, fluent in four languages, desired to hurl profanity at Clay in each and every one of them, but couldn't form the words. He only heard his own voice as Darcy and Clay offered platitudes. *Don't let them make you cry.*

Chapter 6 The Heir Apparent

"I have had it!" shouted Adam. "If you don't get it together when we resume, you're gone!"

The dancers grumbled away, avoiding all eye contact with Adam.

"Adam, is everything okay?" asked Lairea.

"You saw that lousy uprock number. I've seen more artistic conflict in a bar room brawl."

Lairea chuckled. "That's why I always ask you back. You're a perfectionist."

Adam looked at his phone. Clay had called for the eighth time that morning.

Dancers filed back into the rehearsal room and they spent the remainder of the day working hard while Adam barked orders at them. Offstage, Clay waited by a piano.

"Will you please talk to me?" asked Clay.

Adam pursed his lips.

"This whole thing with Darcy is nothing more than a means to an end. It's business. That's all."

"Monkey business is more like it," said Adam.

"I take no pleasure in any of this," added Clay. "That goes for Darcy, too. She was only doing her job."

"Funny, I thought she was your business manager. So exactly which job are we talking about here? Hand job? Blow job? I know you said the label has been pressuring you to marry, but your dedication to go above and beyond your contractual obligations is outrageous!"

"But that's the thing, Adam. I'm contractually obligated."

"Give me a break! You expect me to believe that you are required to marry your business manager?"

"Once we have a kid Darcy and I will amicably divorce, then you and I can have a fresh start."

Adam sat on the piano bench. "I'd say I've found out more about you in the past few days than I have in all our years together. Now a child? I thought you didn't want a baby, Clay."

"An heir might not be so bad. I think I'd like having one of those someday, a legacy. The spittin' image of me."

"Clay, I've offered all of that to you: marriage, a child, all of it and more! Why can't you have a child with me?"

"Because, Adam, you're a man."

Chapter 7 Let's Dance

Columbus, Cincinnati, Cleveland. It didn't matter which city the tour performed in, to Adam each was the same. Thousands of faces, but none wore the smile he longed to see. Work became his loving mistress. She wanted only his body. She asked nothing from him beyond his physical energy, but he always gave much more.

Every kick and stretch was an outlet for the pain of his rejection and loneliness. His extraordinary flexibility displayed on stage made him larger than life.

By the end of the tour, his diminished appetite and intensity with which he danced started to take a toll. He exited the showers as the rest of the dance cast wandered off to a celebratory night on the town. He stalled before dressing to look at himself in the mirror.

Adam was always very fit and lean, but his reflection revealed a gaunt image he'd never seen before. His cheekbones threatened to protrude through his skin. Deep pockets of wrinkles hid the sultry chestnut brown eyes he'd inherited from his Syrian mother. He was strong and yet, in some ways, quite frail.

"Wow! You're ripped," observed one of the chorus line. "Keto?"

"No. Heartbreak diet."

"We're doing drinks at Vinny's on Ninth, Adam. You coming?"

"No. You go on ahead."

"You sure? Someone will snap you up!"

The dancer's voice echoed in Adam's head as he dressed. *Someone will snap you up.*

Odd how the mind worked. Adam noticed his father had called. Upon seeing his father's name, he was reminded how Dr. Safi Khan and Clay were so alike. Both were towering men at the peak of their professions. But unfortunately for Adam, his father never really wanted him either.

Adam returned his father's call.

A pleasant voice answered, "Dr. Khan's office, Nurse Vicki speaking. How may I direct your call?"

"Vicki, it's Adam."

"Your father isn't here. You just missed him. Switzerland conference. Can I help?"

Adam hesitated for a moment. "You worked in genetic engineering. Is that true?"

"Years ago, why do you ask?"

"I need your help, Vicki. I want to become pregnant."

"What? I'm sorry, Adam. This must be a bad connection, let me call you right back."

Adam took a deep breath as he steeled himself in the lighted mirror. "You heard me correctly, Vicki. I want to have Clay's baby."

Chapter 8 The Elusive Dr. Liu

A woman bounced a crying baby on her knee while two toddlers struggled to share a set of blocks at a low plastic table. Adam smiled at them as he made his way to the reception window in the far corner of the clinic. He tapped on the frosted glass barrier, but the shadowy silhouette behind it paid him no mind.

"Give me a minute." More time passed. The toddlers had come to an uneasy truce. Finally, after the day chugged mercilessly forward, a chubby middle-aged woman in blue scrubs and bifocals came to meet him.

"Adam Khan. I have a two o'clock with Dr. Liu."

She clicked at her keyboard, entered a few notes and, without lifting her eyes from the monitor, said, "Have a seat."

Time passed. The day moved to late afternoon.

"I see you waited."

Adam stood, offered a handshake, but was snubbed by the man with salt and pepper hair.

"Good evening, Doctor. My name is Adam Khan. Thank you for seeing me. For a moment there, I thought you were going to bail out on me."

"Follow me."

The doctor took a seat behind his desk, tucking his lab coat under him as he did, and placed a pair of eyeglasses in his breast pocket. "Nurse Vicki told me that you wanted to discuss my genetic engineering work with The Chimera Collaborative."

"Yes."

"That is reason enough for me to want to 'bail out on you', as you say."

"Why?"

Dr. Liu took a deep breath. He stared through a web he made with his fingers while he gathered his thoughts. "I wasn't involved in transplants exclusively. I brokered the manufacture of a pig-human hybrid that served as a source of transplant parts. It was largely responsible for creating life, *human life*, for the sole purpose of destroying it."

"Sounds horrible."

"Yes, well, that is why I choose as I do, running a small clinic to serve at-risk children."

"I had no idea, Dr. Liu."

"You know, Adam, you look a lot like your father."

"I hear that quite often. You know him well, then?"

"I met him some years ago at a renal transplant symposium. Tell me, Adam, does your father know what you hope to accomplish?"

"No."

"I didn't think so. And your partner? Does he know about this?"

Adam was silent and looked away.

"My point is this," continued Dr. Liu. "You are jumping head first into shallow, shark- infested water. The Chimera Collaborative is a group of brilliant, but morally bankrupt, people. The only way to navigate that danger is to be surrounded by people who love you. You will need advocates to tread those waters."

"I'm desperate, Doctor!"

"Why must you make yourself something you are not in order to receive love and affection, Adam? You are a handsome, intelligent young man from an influential and prestigious family. Isn't that enough for you?"

"Sounds like something my mother would say."

"In her conspicuous absence, someone should be the voice of reason."

"You forget, Doctor, my entire family is devout Muslim. I have been pretending to be something I am not for as long as I can remember."

"Your parents don't even know that you're gay?" Dr. Liu emitted a sigh. "This is worse than I thought."

"They know," replied Adam, "but tell themselves I just haven't found the right girl."

"I am sorry. That must be a horrible way to live."

"Will you help me?"

"No, but I will give you information on those who will." Dr. Liu took a zip drive from his desk drawer. "Be careful, Adam. This will be very expensive."

"I make a decent living, and I have a small inheritance," said Adam.

Dr. Liu slid the zip drive to him. “The cost of this endeavor won’t just be in dollars.”

Chapter 9 Change the World

"Your patio suite is just through the garden. The bellman will lead the way. Enjoy your stay in Los Angeles, Mr. Khan."

Adam followed the bellman to the door of his suite. He tipped the man and opened the door to an attractive room. It was serene, with sand-colored carpet and walls. Unlike the severe, modern places he had become accustomed to while traveling, the pale greens and coral accents gave the place a feeling of an upscale, sophisticated home. He kicked off his impeccably polished loafers and placed them in a felt drawstring bag before stretching out on the fainting couch at the foot of the bed.

A knock at the door startled him.

"Enter!"

Clay stood in the open doorway, backlit by the California sun. "Playing hard to get?" His gaze landed on the carpet. "She still isn't pregnant, but wants to adopt."

"I feel bad for you." He shrugged and looked away. "And for Darcy as well, but I still don't like it."

"I'm sorry for all the lies, Adam. Would you excuse me? Gotta use the little boy's room."

Adam took his phone out and pressed record.

"You know what I wish, Clay? I wish that I could take Darcy's place. I wish that we were married."

"Let's not get into that again." Clay bent over to inspect the contents of the mini-bar.

"I wish I had your baby; your biological child."

"We don't live in a world like that, Adam." Clay twisted the cap off a bottle of imported beer and offered one to Adam, who shook his head no.

"What if we did live in a world where all of that was possible? What then?"

"That's not possible, and you know it."

"Forget that. What if it *was* possible? What if we could have all of it? A life together with a child of our own. Would you want it?"

"If circumstances were different, yes. Hey, hold on just a minute. I know that sly look. What are you fixin' to do?"

Adam smiled. "I'm about to change the world."

• • • •

HOURS LATER, WHILE Clay slept, Adam pulled an insulated lock box out from under the bed. The fasteners clicked open, and a ghost of dry ice slithered to the floor. Adam crept to the waste basket, removed a few used condoms, and placed them in the transport cooler.

Flight 2090 from Los Angeles to Osaka Kokusai, confirmed.

Chapter 10 The Devil is in the Details

A twelve-hour, thirty-minute flight to Japan was plenty of time to dream about the life Adam was about to have with Clay. He fantasized about crisp autumn days at Chisholm Moors, he and Clay jumping in piles of leaves with their son. He saw the three of them cuddled under blankets for movie nights at home and the two of them teaching the boy how to ride a bike.

Many hours later, the colossal airbus landed. A limousine driver held a sign with his name printed in English. He spoke a greeting of hello and confirmed his destination.

It was a comfortable drive to the south of Osaka's city center. The destination was an industrial complex with scores of three- to four-story buildings.

"Tochaku shumashita," the driver intoned. "We have arrived. Welcome to Pig Island."

"Pig Island?" Adam worried out the window. "I thought we were going to Sakishima Cosmosquare."

"Hai. Sakishima Cosmosquare. Yes. You are correct. Neighboring residents call it Pig Island. They say on warm nights, when all is quiet, you can hear the squealing of pigs carried on the summer breeze."

The driver stopped outside a sunshine-yellow building with glass and chrome accents. The sign above the series of matching buildings was emblazoned with Kanji characters Adam could not understand, but below in English it read: *Sutherland Medical Partners World Headquarters.*

"Hoteru wa chikaku ni arimasu," said the driver. "The hotel is nearby. I will take your things there, and check you in. Then I will be back to retrieve you once your meeting has ended."

The driver opened the door and directed Adam through the main entrance. Signs directed him through an open air atrium to a separate building labeled 'The Chimera Collaborative'. A door of glass slid open, and a beautiful young woman with dark hair greeted him with a smile.

"Mr. Khan. Won't you please come in?"

"Thank you. I am here to see Dr. Shen Duan."

"Yes, Mr. Khan. It will be a few minutes. We are waiting for Dr. Cantowicz, one of Doctor's most prestigious colleagues, to join the meeting. May I offer you some refreshment after your long flight?"

Adam relaxed onto a curvilinear sofa of the whitest leather.

Opaque glass walls filtered out bright sunshine. Two men in lab coats stood as he entered the room. A glass-sided box rested on a steel cart to the left of a long credenza. In the box, five fuzzy piglets snorted happily.

"Good day, Mr. Khan." Dr. Duan bowed with reverence. Dr. Cantowicz offered a handshake. "I trust you had a pleasant flight?" asked Duan.

"Yes, thank you."

Dr. Duan wore a sly smile. This," he motioned to his sleek modern surroundings, "is my new lab, and Japan my adopted country. Speaking of which, thank you for the timely wire transfer from your private bank to The Chimera Collaborative." Again, Duan bowed.

Dr. Cantowicz motioned toward the case of piglets. "Do you see anything different in these animals?" he asked as he reached into the box with gloved hands, pulling out two squirming piglets.

"No, I don't think so." Adam shrugged. "Their spots are different, and the color of their collars. Nothing else. Why do you ask?"

"The piglet in Dr. Cantowicz's right hand was produced from naturally occurring sex cells, from a male and female pig. The other was made by engineering skin cells into ova. Both specimens are quite healthy, I assure you. They serve as an example of what we here at The Chimera Collaborative can do for you," Duan told him, bowing.

"In this meeting, today," said Dr. Cantowicz, "we will outline your procedures and see to any of your concerns or questions."

"Procedures?" asked Adam.

"Ahh, yes." Dr. Duan resumed his charge of the meeting. "Your situation is quite complex. There are many steps involved. That is why you will stay in close comfort at the local hotel. We keep a large number of suites there, precisely for this purpose."

"How complex can it be?" asked Adam. "Didn't a man give birth years ago?"

"Yes. A keen memory you have, Mr. Khan," said Dr. Duan. "But in that case the patient was a trans-man, meaning he was female at birth. That man had the pelvic construction of a woman. You, Mr. Khan, do not."

"I see, but you and your team can install a functioning uterus as we discussed, correct?"

"Correct," said Dr. Duan. "Our good Dr. Cantowicz is the finest transplant surgeon in all the world."

"Mr. Khan," asked Dr. Cantowicz as he slicked his wavy silver gray hair back with his hand, "am I to understand it correctly that you do not wish to fully transition to a female?"

"No, I don't. I like being a man, but—"

"You wish to experience childbirth?" interrupted Dr. Cantowicz.

"Yes."

"That adds to the complexity of your case. The uterus we transplant must have a thick lining of at least six millimeters to accept an embryo. Normally, we would like to test the new uterus in its ability to function properly, but with no vagina, there is no ability to shed the lining," said Dr. Liu.

"You can still do it, right?"

Dr. Duan nodded. "Complex, yes, but not impossible."

"Doctors, it is more than some perverse thrill that I seek. I love Clay, and he loves me. We both want a biological child, and I want to have that child for him." Adam swallowed hard in correcting his words. "I mean, for us."

"Then why is your partner absent from a meeting that will change his life?" asked Dr. Cantowicz, hands on his hips.

"I know what you are thinking." Adam reached into his breast pocket for his phone. "I have Clay's consent. Listen."

"Yes, I want to have a baby with you, Adam," said Clay's recorded voice on playback. "There's more of the conversation, where he agrees to this, if you care to hear it."

Dr. Duan and Dr. Cantowicz whispered amongst themselves for a moment.

"The cryogenic package of your partner's semen arrived a few days ago. We have much healthy product to use, thanks to your speedy shipment," said Dr. Duan. He waved his hand, and a three-dimensional image appeared before a frosted wall of windows. "This briefly explains the steps we must take to help you and your partner deliver a strong, healthy baby."

The doctors described the hormone therapy that was the required first step, the transformation of Adam's skin cells into eggs, the uterine transplant, and finally the transfer of the in-vitro fertilized eggs into the new uterus.

"Have you any questions?" asked Dr. Duan.

"The uterine transplant… how long it will take to get one?"

"A uterus has been engineered for you from your DNA, tailored just for you," replied Dr. Duan. "As you see, Mr. Khan, the donor and the tissue you have need of will be very much alive."

"Alive?"

"The donor uterus and arteries to feed it will come from a pig-human hybrid. A Chimera," said Dr. Cantowicz.

A cold shudder emanated from the base of Adam's skull to his toes. He remembered the words of the limo driver: *"Welcome to Pig Island."*

Chapter 11 'Roid Rage

"Room 444, sir. Your belongings are already there, waiting for you," said the desk clerk as he coupled the door's lock with an app on Adam's phone.

Testosterone was already coursing through Adam's veins from the anti-androgens administered to him. He danced around his hotel room in a tap-swing sequence from an old Hollywood film. His clothes were hung, organized by color and fabric, and positioned on the rack to ensure they wouldn't touch.

Exhaustion should have overtaken him after the long flight and eventful day of being poked and prodded by The Chimera Collaborative staff, but he was wide awake and famished. He labored over a courtesy tray of mochi that he found both sticky and hard to chew, downing the sweet treats with tea. He felt a little better, but still needed to blow off some steam before treating himself to the large meal he craved.

He decided to go for a run, but on his way to the door stopped to answer an incoming call from his mother.

"Oh, Praise Allah," she said. "You finally answer my call! I was sick with worry. It has been days since I have had word of you!"

"I'm sorry."

"You tear out my heart and all you say is sorry? What if you were injured! Lying dead in the street for all I know! I can tell you are not taking proper care of yourself. What you need is a good woman to look after you. Why can't you find a nice Muslim girl? Then I won't worry so much."

Adam paced during the inquisition.

"I promise I'll eat, but I have to go now."

"Go? Where do you go? I don't like this work schedule of yours. If only you'd listened to your father, and gone into medicine."

"Mom, I have to go."

The paved trail that paralleled the water's edge was scenic and filled with a variety people, even given the evening hour. The multitude of ethnicities confirmed that this place was an international business hub. Clever lighting dotted the local foliage and kept Adam on the path. As he jogged back to his hotel and approached the main door, he thought he heard the sound of squealing pig.

"Did you hear that?"

"Pardon me, sir?" asked the puzzled doorman.

Embarrassed, Adam wiped the sweat from his brow with a sports towel. "Oh, nothing. Long trip."

"We have a most excellent restaurant on the premises. Shall I have the concierge reserve a table for you?"

"Yes. Thank you. I would like that."

A hot shower and full belly worked like magic. Adam was able to get a night of rest. The next morning he rose with the same untamed vigor, dressed, and took another lap around the waterside trail. When he returned and showered, thoughts of Clay plagued him. He checked the clock, calculated Central Time and placed a call, but received no answer.

At his hotel room's door was a male nurse from The Chimera Collaborative.

"Please," Adam motioned for him to come inside. "I was told to expect you."

The man bowed then went straight to work, documenting Adam's vitals. "How are you feeling today?" he asked in flawless English.

"I feel like I'm ready to climb the walls."

"You should feel normal in a few days, but expect a significant drop in libido, appetite, and energy level when you reach castrate levels in a couple of weeks."

"Castrate?" questioned Adam.

The nurse smiled. "No need to worry. Your genitals will not be surgically removed. Once you reach chemical castration, Dr. Cantowicz and his team can proceed with the transplant."

"Dr. Cantowicz— do you know him well? Can I trust him?"

"I would trust no other. He is the finest in all the world." The nurse packed up his instruments and walked to the door. "Remember, you have an appointment back at the campus this afternoon."

The door closed behind the nurse. Adam made a bee-line for his phone.

"Eight, nine, ten, eleven. Come on, Clay, pick up! What has you so busy at this hour?" Adam took a seat on the edge of the bed. A wave of panic crashed over him. He redialed several times until a breathless Clay answered.

"Chisholm here. Talk to me."

"Talk dirty to me is more like it," said an unfamiliar voice in the background.

"Clay? What the hell is up with you? Why are you so winded?"

"You caught me at the Y. My old buddy's in town. Basketball."

"Basketball? Since when do you play basketball?"

"I played in high school." Clay huffed into the receiver.

"And you're playing now?" asked Adam.

"Of course now. What's the trouble?"

"I am barely out of the country 48 hours! You lying, cheating, son of a bitch!"

Chapter 12 Eggs-Actly

"Hold steady. We're almost through."

Adam winced. "Tell me again why you can't just remove some surface skin?"

The technician stopped. "Dr. Duan will be with you in a few minutes. You may dress now."

Minutes later...

Dr. Duan entered the lab. "I am told you have questions, yes?" He bowed.

Adam sat up, trying not to put pressure on his hip and minor surgery. "I have questions about this whole process, Dr. Duan. The purpose of this visit is to start the egg-making process, right?"

"A thoughtful question, Mr. Khan. Skin cells are what we call somatic cells. Meaning, they are non-reproductive cells."

"But the goal here is for me to reproduce."

"Of course, Mr. Khan, but in order to achieve our goal we need sperm, which has been supplied by your life partner, and ova, which is to be supplied by you."

"Makes sense." *I guess.*

"Yes, Mr. Khan, it does. We use somatic cells from you because, as a human born male, your reproductive cells give us more sperm."

"Why my hip? Couldn't you just take some skin cells from the surface?"

"Human skin has many layers, as I am sure you are aware. The surface cells, or epidermis cells, are poor candidates to become specialized cells, as they are damaged by the sun and detergents used to bathe. The cells we just collected will become a pluripotent stem cell. These pluripotent cells can become the specialized cells, or ova, that we desire."

"How do you convince my cells to become eggs?"

"Hormones from ovarian tissue coax the cells to change."

"But I'm a man. I don't have any ovarian tissue."

Dr. Duan clenched his jaw, exposing a cunning grin. "The pig-human hybrid Chimera providing the uterus has ovarian tissue, Mr. Khan."

Adam jumped from the table and rushed to the door. He fled down a long hallway, crashed headfirst into a man wheeling a stainless steel cart, and spun around to shout back to Dr. Duan, "I want a human child! Do you hear me? Human! I won't birth a pig!"

Chapter 13 Chemical Castration

Adam had no desire. The boundless energy he both suffered from and enjoyed from the first phase of anti-androgens left him deflated and exhausted. Staring at the hotel room's ceiling he asked himself, *Why the hell am I doing this?*

Adam's phone rang. He felt around the bed sheets for it.

"Don't hang up," said Clay. His smiling face lit up the screen.

"I'm waiting for an apology," said Adam.

"I'm sorry I made you mad."

"What kind of apology is that? You hurt me deeply, and you know it."

"I'm sorry I hurt your feelings." Clay smiled.

"You're not sorry for what you did."

"I didn't call to fight. I called because I had to tell you that I love you."

The more Adam looked at Clay, the more the dam of ice around his heart began to melt. At first with a trickle, but then mighty chunks began to fall away.

Chapter 14 The Million-Dollar Question

"Mr. Khan, it's a pleasure to see you. And you look so well," soothed Dr. Duan. "You gave us a great deal of worry."

Dr. Cantowicz nodded in agreement, his arms folded across his chest.

"Doctors, I'm afraid I owe you and your staff, especially the man I knocked over, an apology."

"No. Not another word. I blame the heightened levels of testosterone you were experiencing. Your blood work indicated that the spike you endured was well beyond the norm. Fear and uncertainty only fueled the hormone-induced aggression."

Adam shrugged. "Thank you for understanding."

The doctors shared a knowing glance between them, a look that seemed to convey a private message only they understood.

"Allow me to congratulate you on behalf of the entire staff of The Chimera Collaborative in your wise decision," said Dr. Duan. "We will happily move forward. There is, however, a payment that must be addressed."

"In order to proceed," interrupted Dr. Cantowicz. "We'll need another payment of one million U.S. dollars.

"You are asking for the balance?"

"No. This is in addition to the balance you will owe when you deliver."

"What? I signed the papers. We had an agreement!"

"That was before you flew into a blind rage and wrecked the place," Cantowicz sneered.

Dr. Duan walked to a pair of white leather chairs and took a seat behind a massive glass desk. "What the doctor is trying to say is this: It is a complex thing we do for you. All of it is custom-made. The uterus and ovum are personalized for your specific needs."

Adam contemplated the pig parts that would enable his pregnancy. He thought of his strict Muslim mother, and what she might think of it all. He let out a half laugh.

"Something funny?" asked Dr. Cantowicz.

"Consider the additional funds insurance," added Dr. Duan, "that you will not discontinue our services."

Adam put his head in his hands. "I can't come by an extra million easily."

"Do you want a child or not?" asked Dr. Cantowicz.

"I do, but my partner is a very famous man. I want the specifics of the birth of his child to be confidential."

"Very well. Our counsel will draw up a non-disclosure agreement."

Chapter 15 In the Clear

As he scrubbed Dr. Cantowicz said, "You've all been briefed. Any last-minute questions?" His surgical staff was silent as they snapped their gloves and positioned their masks. "Good. I want a clean, fast procedure with zero mistakes."

One by one, the staff filtered into the operating room.

A mask was put over Adam's face.

"Count backwards from ten," said a nurse.

"He's out," reported the anesthesiologist.

"A little music to clear my head and help me focus," said Dr. Cantowicz.

Hours passed.

"He's lucky to have me! Look at the tortuosity of this abdominal artery!" commented Dr. Cantowicz. "No one else would be able to attach the donor uterine artery to this twisted mess."

"We aren't implanting the embryos?" asked a nurse.

Dr. Cantowicz's eyes narrowed as he continued to close. "Not that it's any of your business, Nurse, but no."

"Why not?" she pressed him.

"Don't question me!" he shouted. "The patient just had a major procedure. His sutures need time to heal and the anti-rejection drugs to be out of his system before he can handle the burden of a growing fetus. You of all people should understand that! Besides, his check hasn't cleared yet."

"He will be okay, won't he?" she asked.

Dr. Cantowicz shrugged. "He is a man who just had a uterus implanted. Would anyone be okay after that?"

Chapter 16 Who's Counting?

"Mr. Khan, the doctors will see you now."

Adam walked into an exam room.

"Before we proceed with the fertilized egg transfer, Adam, Dr. Cantowicz and I would like to take a visual inventory of your general health."

Adam lay back on the table.

"The incision is your best by far, Dr. Cantowicz. The delicacy with which you traced the pectineal line of the pubis is genius," said Dr. Duan.

"I was amazed by how good the scar looks," complimented Adam. "I'll be swimsuit- ready in no time, but it's this paunch I can't get over."

"The Chimera uterus has added an inch and a half to your waistline. The extra fat? Not to worry, Mr. Khan. It will help you through the pregnancy," said Dr. Duan.

"I'm the best there is, Mr. Khan. I can fix anything." Dr. Cantowicz admired his own work. "Including that paunch, but first let's get you through gestation."

"We may proceed as we discussed. We will insert the fertilized eggs arthroscopically," said Duan.

"Eggs?" asked Adam, shocked by the word.

"Of the 43 successfully engineered ova, only two survived. One male. One female. We will implant both," said Dr. Duan.

"No!" shouted Adam. "Go with the male one!"

"Mr. Khan, this is a difficult thing you ask of us. We want only to give you the best chance of a productive full term pregnancy. Better odds with two embryos." Dr. Duan smiled.

"This will be the first male pregnancy completed with Chimera ovum and uterus," commented Dr. Cantowicz. "I may never have this chance again!"

"What do you want, Mr. Khan?" asked Duan.

"I want Clay to be happy. I want to give him a son."

"As you wish." Dr. Duan bowed with reverence.

Dr. Cantowicz called for a nurse. "Let's get him draped."

Hours later a bright, silver-white light fell on Adam's sleeping face.

A masked Dr. Cantowicz peered down, smiling at the petri dish that once contained two embryos, one male, one female. It was now, completely empty.

Chapter 17 Baby Daddy

With a firm handshake and a slap on the shoulder, Clay greeted Adam with the same decorum he practiced in public. He tossed Adam's luggage into his truck's bed and drove out of sight of the crowd at Nashville BNA.

"Notice anything different about me?" asked Adam. "I have a surprise for you."

Clay's head bobbled back and forth between Adam and the highway. "Different?" He looked at the extended length of seatbelt around Adam's midsection. "That big ol' gut is different, and that's just for starters."

"I think you will come to love this big ol' gut, as you so rudely put it."

Adam rested his head on Clay's shoulder. "Don't you want to know what it is?"

"Makes no sense how you can go to a place where they eat practically nuthin' and come back lookin' like a heifer." Clay's phone rang. "Hold on, I gotta take this."

Adam spent the drive home munching on saltines to curb his morning sickness, and staring out the window while Clay took call after call. Eventually the truck turned down Clay's private drive. He punched in his key code and the iron gates of Chisholm Moors, embellished with the entwined initials CC, glided open.

Bags and totes littered the master suite. Clay dropped them from his shoulders without rhyme or reason.

"Darcy is spending a little down time in Mazatlán with her girls, and Maria has a long weekend off."

He stripped, and sprang into the bath to find Adam already waiting in a thick mist of humidity, a cloud fallen to earth. The visual appraisal of Adam's nude body was worse than Clay had thought. After an embrace of soft, squishy flesh, Clay gagged, turning his head away. The sting of vomit backed up in his throat.

"Aren't you just dying to know?"

Clay choked back the bile, and offered a generous, "Sure, let me have it."

"I'm not fat. I'm pregnant!"

Chapter 18 Crazy

"I came as soon as we landed," said a breathless but tan Darcy, entering through the front door. "Tell me, is it really true? How did this happen? Please tell me that the label hasn't gotten wind of this."

Clay's forehead furrowed. He ran both hands through his tousled hair, knitted his fingers together, and cradled his neck as he paced. "He's preggers, all right. I see him getting bigger every day. As for how it happened? Not the usual way, that's for sure."

"Obviously, but how?"

"A Chinese doctor in Japan took his DNA and merged it with a pig embryo."

Darcy pushed back a headache from jet lag. Her elbow rested on a sofa arm while her hand soothed her forehead.

"Ha! Crazy. Absolutely crazy. Why a pig?"

"Pigs make good human parts. Oh, and by the way, Adam's also the mother."

"Yeah, right. Seriously. Who's the mother?"

"He is! That Chinese doctor made a uterus from a pig and eggs from Adam's skin cells. All of that got implanted in Adam, who is now, through the miracle of modern science, carrying my child."

Darcy giggled and said, "You're joking, right?" Clay paced. "You're not. Wow! Are you high? Do you want to be? Maybe this is a dream and we're both asleep."

"I wish."

Darcy pinched herself. "It makes no sense. How did he get your sperm?"

"Used condoms. Dug 'em out of the trash."

Clay stopped with his pacing and sat on the sofa beside her.

"This pig stuff is weird. I thought Adam was Muslim. They don't even eat pork."

"He observes some traditions, but mostly to please his mother."

"I'm so confused."

He turned to her. "I need your help with this, Darcy. If J.D. gets wind of this, it's the street for all of us."

Chapter 19 Ultrasound of Silence

"The Chimera Collaborative strikes again," said the obstetrician jokingly. "Do you want the gender to be revealed?"

"I already know," said Adam. "It's a boy."

The doctor examined the screen and pressed her scope to Adam's abdomen for a very careful review. "Odd."

"What is?"

"For a moment I thought I heard..." She scrolled through the chart from The Chimera Collaborative, took a deep breath, and offered a calming smile. "You have a normal, healthy baby," she reported, "but it's large for 20 weeks."

"Takes after Clay already," commented Adam.

"Amazing and very strange," commented the obstetrician. "Mr. Khan, you're the first in what will someday be a normal occurrence."

"What is that dark spot on the screen?" asked Adam.

"Uterine tissue. Nothing more."

She stood, excused herself, and exited before Adam could get in another word.

Chapter 20 Wayward Son

Clay took a seat on a big brown leather sofa. His foot tapped to the waiting room music. A spotted cowhide rug softened his boot's enthusiasm. He hummed along with the tune and many others. He soon started to think something was amiss. J.D. never kept him waiting.

"Clay!" J.D. burst into the room, loud and animated. "C'mon in, boy! Make yourself at home. Hope I haven't kept you too long. April, honey?"

"Yes, J.D.?" asked his assistant

"Hold all my calls." J.D. pulled a massive set of weathered barn doors closed behind him. "There. Now we can have ourselves a private conversation." J.D. walked to desk chair built from elk antlers. "How are things, Clay?"

"I've been busy in my downtime, J.D., recording at home. I have some great stuff to add to the second leg of the *American Man* tour."

"Yes. April made sure I got the demos. Fine material, just fine, but that's not why I asked you here today. Son, you and I have been doin' bid'ness a long time, isn't that right?"

"Sure. I figure we've known each other for a good ten years or so."

"That's right." J.D. poured himself three fingers of bourbon. "Would you care for a snort?" He offered Clay a glass.

Clay waved it away.

"I insist. Nothing good comes from drinking alone," said J.D.

Clay took the drink in hand. "What's on your mind, J.D.?"

"Son, I'm afraid we're going to have to let you go."

"Come again?"

"I'm sorry for having to do this, but you are in clear violation of your contract. I have no choice." J.D. took a long, soothing sip. "Isn't that good bourbon? Go on, take a sip."

"Violation? I followed your advice and married Darcy. I held up my end of the bargain."

"I'm not talking about Darcy. I'm talking about that little dancer friend of yours, Adam." J.D. held out an envelope. "Go on. Take it. It don't bite."

Clay removed dozens of photos taken of Adam while under confinement at Chisholm Moors. One of a shirtless Adam getting some air from the master bedroom balcony stood out.

"Look here, son. I know you have exotic tastes. Hell, we all got something, even me, I like 'em young. But you haven't been discreet, boy, and now you're a liability because of it."

"I'm getting the boot for helping a friend?"

"Come on, now. It's me you're talking to, Clay. I know he's more than your friend. I know that he's your pregnant lover. He has a baby bump and titties, for Chrissake."

"How do you know 'bout that?"

He handed Clay a copy of his contract. "Take it home and compare it to your copy. Since it's probably the first time you've read it, have your lawyer explain it. You are in violation of the morality clause on page 6."

"Morality clause?"

"Confirmed!" shouted J.D. with blazing rage. "Son, what you're doin' doesn't only violate your contract, it goes against God, nature, and The United States of America!"

Chapter 21 Replacement Parts

"You're Clay Chisholm," reminded Adam, biting into a piece of dry toast. "Any other label would kill to have you. I say screw the bastards and just move to pop altogether."

"Language!"

"Sorry. You know how I get when I'm angry or upset. What I'm trying to say is that it can't be so bad."

Maria entered. She poured fresh coffee for them, glanced at Adam's ever increasing belly, and made a sign of the cross as she muttered something in Portuguese before bolting to the safety of the kitchen.

Adam taunted her as she hurried away. "You'd think she never saw a pregnant man before."

"Leave her alone," said Clay. "It's like I'm being blackballed or something."

"Someone is at the gate," announced a computerized voice from a wall monitor.

Clay looked at Adam. "I was young and dumb. I thought I could trust J.D. because he was my friend."

Maria peeked around a corner. "Mr. Chisholm, your wife's on her way up. And there's a Christophe Talan in the car behind her."

"Ask your father if he can help with the last of the baby money."

"Clay, there's no way I can do that. My father can't know about this, not ever!"

"You haven't thought about how you're going to explain this?" Clay slammed his fist to the table top. "Damn if we don't have ourselves one fine mess here!" He stood, grabbed his denim jacket, and made his way for the door. "See y'all later."

"Hey, wait! I have an OB appointment."

"Darcy will take you."

"Me? Don't drag me into this freak show!" objected a startled Darcy as she entered the room.

Darcy and Adam followed him to the drive where a young dark-haired man with fine features sat in a shiny red convertible. Clay hopped in without opening the door.

"Huh. Adam, would you look at that," Darcy mused with a flip of her bleached blonde hair. "Looks like you've just been replaced."

Chapter 22 Ob-Fuscate

"102. How long have you had this fever and chills?" asked the obstetrician.

A pasty-gray Adam patted his brow with a linen handkerchief while Darcy sulked in the corner. "This morning at breakfast the nausea and belly pain started. It's that spot, the dark one on the last sonogram. There's something wrong, I just know it."

"I ordered another ultrasound. The tech will be right with you."

As the obstetrician reviewed the results, she scrolled through Dr. Duan's notes from the implant. Without a thought, she switched images.

"Mr. Khan," she said, showing his recent sonogram on a large screen, "as you can see, there is no spot." She enlarged the image so both Adam and Darcy could see. "You have a slight infection. I want you to start a course of antibiotics. The nurse will see you out."

Chapter 23 A Debt Past Due

The obstetrician breathed a sigh of relief at the sound of the door locking behind her. She called the reception desk, cleared her appointments for the rest of the day, and asked that she not be disturbed.

She slid into her ergonomically-designed armchair and pulled up Adam's file. The sonogram presented to Adam from earlier that day was textbook. The fetus was healthy with no signs of abnormality, but there was indeed a mass in the uterus and it was growing. Her sleight of hand was brilliant, but she wondered, *Did he buy it*? The bloodwork, fever, nausea, all of it indicated to her that the mass was the second zygote that failed to attach.

Dr. Duan appeared on her screen from half a world away. He pursed his lips in a poor attempt at a smile. "Good day, Doctor. You wish to speak with me?"

"Yes, Dr. Duan. Thank you for getting back to me so quickly. It's about The Chimera Collaborative patient, Adam Khan. I sent his chart. I trust you have had a moment to review it?"

"Indeed."

"I believe it is the additional embryo that you implanted. Given his inability to shed the necrotic tissue due to his lack of a vagina, he is suffering an infection."

"Did you biopsy the mass?"

"No. The patient was led to believe that there is no mass present, making the need for a biopsy moot. I would not have been able to explain the discrepancy."

"Course of action?" asked Dr. Duan.

"I started him on a round of antibiotics just as I indicated in the chart."

"You will not treat Adam Khan unless I specifically order otherwise."

"He is the world's first pregnant man, Dr. Duan, an asset to this company's reputation at the very least."

"He is months in arrears. He has 48 hours to make this overdue payment."

Chapter 24 A Rude Awakening

Noises of an unruly symphony forced Adam into consciousness. It began with the murmur of a distant rumbling he couldn't quite place. Then the tempo picked up with a call from his mother, which he sent directly to voice mail before tossing his phone to the empty side of his bed. For a few moments he settled into the rhythm of his own breathing, but the crescendo jostled when the gate buzzed and Clay called at the same time.

"I thought you'd never pick up," said Clay.

"Why the hell did you wake me up to buzz your sorry ass in?"

"I can't buzz in. They're everywhere!"

"Who's everywhere?" Adam looked out the window to see protesters at Clay's gate.

"I've called the cops," said Clay. "Stay inside! Better yet, get to the safe room! These people are crazy!"

A little sleepy and feverish, Adam left the bedroom and stumbled to the mansion's private bunker. It had been years since Clay showed him the trick to enter, but even in his delirium he was able to locate the false panel on the library wall and lock himself behind the deadbolt.

Chapter 25 The News

The next day, the obstetrician arrived at her clinic to hear a member of her staff say, "It's all over the news! Clay Chisholm's boyfriend is pregnant!"

The obstetrician hurried to her office and shut the door for added privacy. She opened her browser to find Dr. Duan streaming a worldwide live conference from Osaka.

Chapter 26 Ave Maria

As he sat and waited for Clay to fight his way past the mob, he scrolled through his messages. All of which were from his mother.

Beep!

"Adam, this is your mother. I am very worried. I barely hear from you for months, and now there are terrible rumors going around about a man with your same name. Please call me back."

Beep!

"Adam. There are shirtless pictures of you standing on a balcony. Tell me that this, these things they say, are simply not true. Please say to me that this is a lie!"

Beep!

"There is a news conference on the internet with Dr. Shen Duan, a geneticist in Japan. You were in Japan many months ago. My son, my gift from Allah," she sobbed. "What have you done?"

Beep!

There was a long pause. Adam could only hear his mother's erratic breathing, as if she had just run a marathon. "Pig parts?" her voice cracked. "Swine? You have defiled yourself with the unclean to have the child of your homosexual lover?" She sniffed, cleared her throat. Her voice was low and calm. "You are dead to me."

Darcy entered the bunker to find Adam alone in a corner chair.

Adam swallowed hard and took a few gasps to help calm down. His face was blotched with grief. "It's my mom. She knows now—about this, I mean. News of my condition has spread all over the world."

Darcy eased down beside him. "She is a very religious woman. What else did you expect? Give her time. Who knows? Maybe she'll come around."

He blew his nose into a tissue.

"Are you feeling any better?"

"The nausea has subsided a little, and there are no more sharp pains in my abdomen. There's only a dull ache now."

She made her way for the door. "Give me a minute, okay?"

Adam sank into his chair and dozed off in the muffled quiet.

Darcy ran into Clay in the hallway. "It's about time you showed up! I need to talk to you. He's running a fever."

"That's not so bad, is it?"

"I don't know, I'm not a doctor! He was prescribed antibiotics, but if you give a hoot about that baby you'd better get him some medical attention."

"We got ourselves an angry mob out there." Clay braced himself along the wall at the side of a window. He peered through the layers of silk drapery to see a low-flying drone. His blood boiled. "I'm going to blow that thing out of the sky!"

Darcy grabbed him by the arm. "We have more important things to deal with."

The noise from the deadbolt roused Adam. Clay knelt beside him. "Darcy says you're running a fever."

Rat-a-tat-tat-tat! Rat-a-tat-tat!

Gunshots broke glass upstairs.

"Balcony doors," said Clay. "Big windows on them doors."

Rat-a-tat-tat-tat! Rat-a-tat-tat!

Adam exhaled and rubbed his aching belly. "No rest for the wicked."

Maria ran from the kitchen toward the safe room. "I heard gunshots! Is everyone all right?" she asked.

"We have to get outta here!" Clay exclaimed.

"Don't worry, Mr. Chisholm," said Maria. "I know what to do."

Chapter 27 Parcel Pick-up

A Zon-Zon truck blared its horn at the mob outside the gate of Chisholm Moors, but the maddened crowds would not disperse.

"What the hell's going on?" asked Harry.

"Back that truck up, you crazy bastard!" ordered a police officer.

"I got a delivery for Mr. Clay Chisholm. Here's my manifest. See for yourself."

The officer eyed the paperwork and walked off. After several minutes of enraged screaming and chants of doom, the officer returned.

"What's going on here with all the signs?" asked Harry. "I don't get all of this Satan's baby stuff."

"The King of Pop Country's gay lover is pregnant."

"Yeah, so why is that such a big deal?"

"The pregnant one is a man."

Harry drove his delivery van to the rear service entrance of Chisholm Moors. He backed into a cargo bay with a small loading dock. Maria met him.

"Thanks so much for doing this," she said.

"Corporate will have my head, Maria, but I'm happy to help a friend."

She opened the door for him that led to a utility room and a catering kitchen with connecting butler's pantry.

"I talked to the cop at the gate. This is for real, huh? Just like you said?"

Maria made the sign of the cross. "Oh, yeah. It's real, all right."

They heard footsteps approaching. Clay was carrying Adam, and Darcy was weighed down with countless pieces of luggage. Harry hurried to relieve her of the bags.

"I want out of here. This is too crazy for me!" said Darcy.

"I need you here," replied Clay.

"You owe me big!" She pointed a long red fingernail at Clay.

Adam winced with a tremor of pain in his abdomen.

"We'd better hurry," said Harry.

Clay propped Adam up with pillows and covered him with blankets.

Harry took his place in the driver's seat.

Chapter 28 Accolades

The day wore on with one patient after the other, but none mattered more than the one the obstetrician lost. It turned out that Dr. Duan would have the all of the fame and glory that he planned for himself, despite the non-disclosure agreement Adam requested. Coverage of his amazing feat was everywhere. She kicked her shoes off and turned up the sound on her computer.

"Chinese geneticist Dr. Shen Duan stuns the world by surgically enabling the pregnancy of a genetically born male!"

She opened the door of the credenza behind her desk. Seething with jealousy, she reached for a packet of Vitamin C powder and emptied it into a rocks glass along with a healthy double shot of vodka. The vitamin powder fizzed to life. She stirred it, collapsed into her chair, and took a sip.

Is there no end to this? she wondered.

"Tell us, Doctor, since the patient did not fully transition to a woman and has no vagina, how will your Chimera baby be born?

"Quite simply. A cesarean delivery is planned. As you know, it is a very common procedure."

"A common procedure that was to have been MY PROCEDURE!" she yelled at the screen.

Chapter 29 Dr. Liu's Lament

"Get the door for me will ya, Harry?" asked Clay, his arms burdened with a fevered and delirious Adam.

They burst into the lobby of a familiar clinic. The full waiting room stared at the dramatic spectacle unfolding before them. Young children, bewildered by the scene, either clung to their mothers' pantlegs or motioned to be held.

"We're here to see Dr. Liu," said Clay.

The receptionist opened the door to the exam room without delay.

Dr. Liu entered and washed his hands. He drew a pair of gloves over his hands and immediately went to work. "I wish you had taken him to an emergency facility rather than travel so great a distance. Nurse, get him started on 500 milligrams of erythromycin and mycophenolate. We're going to have to do an emergency C-section."

The nurse started an IV.

"Hey now, is that stuff safe for my baby?" asked Clay.

"Erythromycin has been around for a long time. It has a history of safety in pregnant patients," said Dr. Liu. "Mycophenolate is newer but safe for short term use, as we're doing here. The goal is to stabilize Adam the best we can before the procedure."

Dr. Liu motioned Clay to the side, away from Adam. "The severity of his symptoms indicate to me that this is a case of organ rejection. I have ordered an ultrasound because I haven't seen his records, but I wonder why he wasn't treated with steroids to address it. Adam is 36-weeks' pregnant. The life of Adam and the fetus are at risk."

Hours passed. Clay paced the waiting room. Dr. Liu suddenly reappeared.

"Mr. Chisholm, Adam is resting. The baby is stressed, but stable. I have detected an echo in its heartbeat."

"An echo?"

"It may be as simple as a newly developed arrhythmia. No need to worry. We will know more when the sonogram is available. I have assembled my colleagues. An anesthetist is already onsite, and a nurse who has a great deal of experience with Chimera-grown organs is on the way."

"Doctor," a woman burst in on them and interrupted, "I'm Jocelyn Smythe, from Sutherland Medical Partners LLP. I was Adam's attending obstetrician in Nashville. I'll scrub in if you'll have me."

Chapter 30 The Last Word

Clay sat in the waiting room. The clinic door opened.

"I demand to see my son!" shouted a woman, followed by a man. "We're Doctor and Mrs. Safi Khan. We have been told our son, Adam Khan, is here. What of his condition? How is he?"

"Dr. Liu will be out to talk with you when he is able. Have a seat."

When they turned to survey their surroundings they found Clay standing before them, wearing a wide, toothy grin.

"Dr. and Mrs. Khan. I know we've never met, but I'd like to introduce myself. I'm Adam's partner, Clay Chisholm."

Dr. Khan returned the handshake out of habit.

Adam's mother's face contorted at the gesture. Her posture became rigid. Her amber eyes burned them both. "You shake his hand?"

"I know I'm a stranger," said Clay, "but since we have our love for Adam in common..."

"You dare speak to me about your sick perversions and call it love?" said Adam's mother. "You, who have led him away from righteousness? Be gone from my sight!"

"Now hold on there, Mrs. Khan. I know you're upset and all. We are all concerned for Adam, but you got it all wrong here about me and him."

Her brow furrowed. "Concerned for Adam? Were you concerned when you lured him from me? When you turned him from all that is decent?"

"Jamal, please," said Adam's father. "Let's go and sit." Dr. Khan led his wife by the elbow to a pair of chairs.

"Unhand me, Safi!"

"Jamal, please. Tensions are high. Let it pass."

"I blame you for this!" she shouted.

"Jamal, lower your voice. Do not make a scene."

"You made me raise our son is this amoral place, Safi. Why?"

"My work and our lives are here."

Mrs. Khan's voice rose until it was no longer a whisper. "And now we have a critically ill son and some filthy sodomite who thinks he has a claim to our grandchild." Mrs. Khan leapt from her chair, racing across the room until her face was just inches from Clay's.

"Jamal!" Dr. Khan stood. "Come sit down! This solves nothing."

"You are the father?" she seethed. "But you are a married man, no? Married to some Banks woman, is that not correct?"

"Yes, but..."

"So, you are a homosexual and an adulterer?"

"My relationship with Darcy is business."

"Adam owns his own home, a condominium in the city."

"He likes his own space when I'm travelling."

"Have you paid a dime toward any of his medical alterations?"

"Well no, but..."

"Yes, but! No, but! But, but, but! And we are to believe that you are our grandchild's father just because you say so?" she questioned, sneering.

Dr. Khan remained seated, his head buried in his hands.

Clay flushed red with anger. The vein in his neck throbbed. "Adam took my used condoms and shipped them to Japan! I didn't know!"

Mrs. Khan had the last word. She spit in his face.

Chapter 31 Lost in the Code

"We only have a few minutes," announced Dr. Liu. "Nurse Vicki has arrived and is scrubbing in. The patient is prepped and ready for surgery."

The ultrasound tech entered, a worried look on her face. "Doctor, I think you'll want to see this."

There was a crashing of carts and the noise of organized confusion in the hall. "DR. LIU! COME QUICK!" the anesthesiologist yelled. "He's coding!"

"There's no time," said Liu as he and Jocelyn rushed past the tech to the operating room.

Nurse Vicki was administering CPR.

"Where's the cart?" asked Dr. Liu. The AED rolled in and was brought online. "Clear?"

"Clear!" shouted Nurse Vicki as Dr. Liu pushed the metallic paddles to Adam's chest.

Adam's body seized with the first shock, but there was no heartbeat.

"Clear?"

"Clear!" responded Vicki

Again, no pulse.

"Clear?"

"Clear!" responded Vicki.

"What do we have? What do we have? Come on!"

"Clear?"

"Clear!"

"Come on. Come on."

"Clear?"

———^——^——^

"Hold!"

"Good boy," said Dr. Liu.

"Nurse," said Jocelyn, "help me get him onto his left side." Jocelyn held her stethoscope to Adam's stomach.

"Fetal pressure dropping!" yelled Nurse Vicki.

"Did he have a DNR?" asked Dr. Liu.

"No," answered Jocelyn.

"We have to go in now!" shouted Dr. Liu.

Dr. Liu ran to the reception room. The startled trio rose to meet the urgency.

"He was in cardiac arrest," said Dr. Liu.

"No! My son!" cried Mrs. Khan.

"I can scrub in," said Dr. Khan.

"No. I have a capable team. Besides, I need you here with Mrs. Khan. The situation is dire. Adam has a pulse, but it is irregular and his BP is low. The fetus is showing signs of stress. Adam had no DNR. I need an answer, and I need it now."

"Save my son!" shouted Mrs. Khan.

"What about *my* son?" objected Clay. "That's my boy in there!"

"I'm sorry, Mr. Chisholm, but you and Adam are not married. The Khans are Adam's next of kin. The decision is theirs." Dr. Liu hurried back to the surgical suite.

"BP is dropping!" said Vicki.

Dr. Liu pushed forward, making himself a spot to better gauge the ongoing procedure. "The Khans have asked that Adam's life take priority."

"I'm going for the baby," said Jocelyn.

"Smythe," ordered Dr. Liu, "I won't have any insubordination from you! The Khans' decision stands."

Without pausing, Jocelyn reached inside Adam and pulled the baby out through the incision. A perfect baby boy. She then reached back in for the second mass that she'd kept hidden from Adam. The baseball-sized life squirmed as it emerged.

SQUE-E-E-AL! SQUEAL!

Nurse Vicki's eyes bulged in disbelief. "Oh my God, nooo!"

"I have a bleeder. The abdominal artery has ruptured!" shouted Jocelyn.

"We're losing him! Smythe, this artery is highly irregular! I can't stop the bleeding! I need an assist! NOW!" shouted Dr. Liu.

"He's still flat-lining, Doctor," said Vicki.

Dr. Liu turned to Jocelyn. "I'll have your license for this!"

"Go right ahead!" yelled Jocelyn. "You yourself said that the abdominal artery was highly irregular!"

She looked around the room at the others glaring back at her. "He did say that."

Wearing better than two liters of Adam's blood Dr. Liu persisted, even though he knew his efforts were futile. He'd promised the Khans he would save their son. He failed to save Adam on this day, and months earlier he'd failed to save Adam from himself. He dropped his instruments and hung his head.

"14:43," Jocelyn stated, noting the official time of death. She gave Dr. Liu a sideways glance she finished her Apgar assessment of the infant twins before drawing the cord blood. "I think we have a case of skeletal dysplasia in the female."

Nurse Vicki took the infants' measurements and administered shots of vitamin K. "Oh, little girl, you didn't ask for any of this."

Chapter 32 Double Take

Dr. Liu showered and donned a clean set of scrubs. He entered the reception area. The Khans and Clay rushed to him.

"How is our son?"

"My boy...he's okay, right?"

"I think it is best if you all sat down."

"Why? What is the matter? Safi? Make him tell us," said Mrs. Khan.

They flocked to their chairs, but were hardly at ease.

"Dr. and Mrs. Khan, I respected your instructions to save Adam at all costs."

"NO! This ain't right!" Clay stood, snarling.

Dr. and Mrs. Khan breathed a sigh of relief.

"Please, all of you, let me finish. Unfortunately, we discovered a severe genetic abnormality in Adam's abdominal artery. Even with the precision that it was joined with the Chimera uterine artery, it burst. We were unable to save him."

"What? Noooo!" Mrs. Khan wailed. "This cannot be true. We instructed you to save Adam." She began hitting and flailing at Dr. Khan. "My only child! My most beloved is dead because you sit here, like a lump! You are no surgeon! What kind of man sits while his child dies?"

Dr. Khan stood. "I have to see him!"

"No, Dr. Khan. I know that you deal with the realities of surgery every day, but it is different when it is one of your family. Please sit."

"No! I must see my son!" Dr. Khan pushed past Dr. Liu and ran to the surgical suite, opening door after door until he found Adam's bloody corpse on the table.

"AAAAAGHHH!" Dr. Khan's cries of grief were so loud, they could be heard by Clay and Mrs. Khan back in the reception room. "I must see this abdominal artery."

Dr. Liu led him to the body, handed him a pair of gloves and safety goggles, and pulled back the sheet that covered Adam's open mid-section.

Dr. Khan stood, weeping softly as he examined the Chimera uterus and the blown artery. After several minutes, he turned away. "I have seen enough."

Nurse Vicki took Dr. Khan's eyewear and gloves. He went back to the reception room where Clay and Mrs. Khan stared holes into him, but eventually he was able to muster the words. "I saw Adam's body, and the abdominal artery. It is as Dr. Liu says."

Tears streamed down Clay's cheeks. Mrs. Khan dropped to her knees, and with palms to the floor she began to pray.

"What about my boy?" Clay sniffed.

"Your twins are doing fine, Mr. Chisholm."

"What? Twins? This is no time to be joshin' me."

"Dr. and Mrs. Khan, I'd like to introduce you to your grandchildren. If all three of you would like to follow me please." He led them to an unused surgery where two bassinets were placed side by side. Nurse

Vicki wheeled the first one up for them to get a closer look.

"This is the male child," said Dr. Liu. "Eight pounds three ounces, Twenty three inches in length."

"Would ya look at that? A regular chip off the ol' block! He even has my hair!"

Dr. Khan held his wife close as she wiped her eyes.

"And this is the female infant." Jocelyn wheeled the second bassinet toward them.

"She is quite small, birthweight is one pound, six ounces."

Mrs. Khan shrieked at first glance. "What is that thing? You!" She poked her finger in Safi's chest. "You bring us to Manhattan, this Pig Island, and this is what we get for it!"

"Small? That's all you've got to say? It's more than small! Why, that thing ain't even human."

"I assure you, she is very human, a pig-human hybrid Chimera. She is also your daughter."

"Oh, no you don't! That right there, that boy, that's mine! He even looks like me! I got nothing to do with this monster."

"Do not speak to me of this abomination!" Once again, Mrs. Khan was filled with anguish. Her beautiful reddish brown eyes were bloodshot. "We are taking our grandson home!"

"Both children are Adam's, and are your responsibility," Dr. Liu interrupted.

"You're not taking my son anywhere!" Clay bellowed.

"Try proving paternity in court. We will fight you, you glorified street brusquer!" Mrs. Khan yelled.

The shouting started the male child to cry as the argument spilled into the hallway. The female infant, agitated, took a cue from her brother as Dr. Smythe wheeled her away.

SQUEAAAAAL! SQUEAAAAL!

Chapter 33 Gone Baby Gone

The battle royale raged on between Clay and his would-be in-laws. The body of Adam was still warm from the remnants of life, yet they no longer seemed to mourn him.

Mrs. Khan shouted something past her anger, aiming it at Clay. She then bit the area between her thumb and forefinger, with palm upraised. Dr. Khan's eyes bulged at the gesture.

"There's no need to insult my manhood," said Clay. "That's right. I know what you said, and what that hand biting thing means. Your son cursed a blue streak, so yeah, I've picked up a few things. I'm not your sister, by the way."

"We shall see!" Mrs. Khan lunged for Clay's throat. The argument intense, and bordering on physical violence. Dr. Liu called for the coroner and the police.

Vicki went back to the surgical suite to prepare for the body pick-up. "Dr. Smythe? The Khans declined an autopsy, citing religious reasons. The bus is here. Dr. Smythe?" She looked around the clinic, searching room after room, but the OB was nowhere to be found. Smythe's locker had been cleaned out.

As the EMTs bagged Adam's body, Nurse Vicki noticed something. "What? Where the heck is the uterus?" Overcome with panic, she ran as fast as she could to the room where the infants awaited their hand- and footprints. One bassinet was conspicuously empty.

"Dr. Liu! Come quick! The Chimera! The female twin! She's gone!"

If you enjoyed The Pregnancy of Adam Khan. Please leave a review!

Are you curious about what else The Chimera Collaborative is up to? Pick up a copy of Zhu. In it, a sick girl befriends her Chimera organ donor, a pig-human hybrid. https://books2read.com/u/barGZ6

Other Work by J. R. Brienza

The Belt of Orion: Nibiru Rising

The Belt of Orion: Ramirez Reveals Planet X

Zhu

Between Heaven and Hiroshima

The Pregnancy of Adam Khan

P.S. Follow me on Twitter @jrbelementals, Facebook @JRBrienza Pinterest @jrbrienza, or on my website at www.jrbrienza.com[1].

1. http://www.jrbrienza.com

Don't miss out!

Visit the website below and you can sign up to receive emails whenever J. R. Brienza publishes a new book. There's no charge and no obligation.

https://books2read.com/r/B-A-VKBL-DUQGB

BOOKS 2 READ

Connecting independent readers to independent writers.

About the Author

At the tender age of 8, J. R. Brienza was inspired by time spent in a cardboard washing machine box to begin her writing career. Her third-grade teacher was a progressive young idealist that encouraged said behavior.

Adulthood landed Brienza in the too-cool-for-school world of the professional creative, where she spent more than twenty years as an interior designer, and Green Advantage certified general contractor.

As bugs often do, the writing bug kept biting. When she isn't penning her next book, or saving the world from shiplap, she spends time in her Pennsylvania home with her longtime love and their pets.

Read more at www.jrbrienza.com.

Milton Keynes UK
Ingram Content Group UK Ltd.
UKHW011558110124
435868UK00001B/15

9 781735 377308